CATERPILLAR THIRTY

PHOTO ARCHIVE
Including Best Thirty, 6G Thirty & R4

Bob LaVoie

Iconografix
Photo Archive Series

Iconografix
PO Box 446
Hudson, Wisconsin 54016 USA

© 1999 by Bob LaVoie

All rights reserved. No part of this work may be reproduced or used in any form by any means... graphic, electronic, or mechanical, including photocopying, recording, taping, or any other information storage and retrieval system... without written permission of the publisher.

The information in this book is true and complete to the best of our knowledge. All recommendations are made without any guarantee on the part of the author or Publisher, who also disclaim any liability incurred in connection with the use of this data or specific details.

We acknowledge that certain words, such as model names and designations, mentioned herein are the property of the trademark holder. We use them for purposes of identification only. This is not an official publication.

Iconografix books are offered at a discount when sold in quantity for promotional use. Businesses or organizations seeking details should write to the Marketing Department, Iconografix, at the above address.

Library of Congress Card Number: 99-71749

ISBN 1-58388-006-2

99 00 01 02 03 04 05 5 4 3 2 1

Printed in the United States of America

Cover and book design by Shawn Glidden

Edited by Dylan Frautschi

Iconografix Inc. exists to preserve history through the publication of notable photographic archives and the list of titles under the Iconografix imprint is constantly growing. Transportation enthusiasts should be on the Iconografix mailing list and are invited to write and ask for a catalog, free of charge.

Authors and editors in the field of transportation history are invited to contact the Editorial Department at Iconografix, Inc., PO Box 446, Hudson, WI 54016. We require a minimum of 120 photographs per subject. We prefer subjects narrow in focus, e.g., a specific model, railroad, or racing venue. Photographs must be of high-quality, suited to large format reproduction.

PREFACE

The histories of machines and mechanical gadgets are contained in the books, journals, correspondence, and personal papers stored in libraries and archives throughout the world. Written in tens of languages, covering thousands of subjects, the stories are recorded in millions of words.

Words are powerful. Yet, the impact of a single image, a photograph or an illustration, often relates more than dozens of pages of text. Fortunately, many of the libraries and archives that house the words also preserve the images.

In the *Photo Archive Series,* Iconografix reproduces photographs and illustrations selected from public and private collections. The images are chosen to tell a story—to capture the character of their subject. Reproduced as found, they are accompanied by the captions made available by the archive.

The Iconografix *Photo Archive Series* is dedicated to young and old alike, the enthusiast, the collector and anyone who, like us, is fascinated by "things" mechanical.

Table of Contents

Introduction ------------------ pp. 7
Best Thirty -------------------- pp. 6, 8-30
S and PS Series Thirty ------ pp. 31-83
Wide Gauge Group ----------- pp. 84-93
6G Thirty --------------------- pp. 94-123
R4 ------------------------------ pp. 124-126

Best Thirty with open radiator sides. Notice the absence of track adjusting springs on the Best machines.

INTRODUCTION

The Best Model Thirty tractor was originally born from replacing its predecessor, the Best Model Twenty-Five of 1919 vintage. The Thirty was meant to compete with the Holt Mfg. Company's 5 Ton tractor, which was its main stay in the mid size track tractor market. The Best company had long used the trade name "Track Layer" to describe its crawler-type machines. During the same period, the Holt company used a trade name of there own, "Caterpillar." When the merger, or perhaps better termed consolidation, of the Holt and Best tractor companies occurred in 1925, the name Caterpillar stayed and became the new name of the company.

The "S" Thirty, by most reports, was truly an exceptional machine as compared to the Holt 5 Ton. Several old timers have related stories saying that the Best tractors were named accordingly since they truly were the best! The Holt tractor gained popularity from its strong dealer network and almost exclusive use as a prime mover during WWI. Very few Best dealerships were found East of the Mississippi while the Holt tractor was becoming popular throughout the country.

The first Model Thirty was known as the Model "S." The first serial number began with S1001. It was a rear seat or tail seat style tractor for use mainly in the orchards of the San Joaquin valley of California. The "S" Model Thirty was of a two speed design and was powered by a 4 3/4 by 6 1/2 inch bore and stroke 4 cylinder engine of the Best design. Its steering was accomplished by two levers that were situated almost horizontally, extending over the right fender. The "S" Thirty was also one of the earliest tractors to have wet steering clutches (clutches that ran in oil) like that of the Holt, and later Caterpillar 2 ton. The wet steering clutch arrangement was changed at serial number S1276 in 1922 to the dry type. Also, at tractor number S1276, the size of the radiator and water pump was increased as well as many other improvements.

The "S" Thirty serial number S1800 was tested at Nebraska in 1923, bearing test number 99. This was a retest from test number 77 that was conducted on tractor serial number S1140. The tractor performed well with its new modifications.

At serial number S2076, other changes took place as well. The changes include a different transmission gearing combination and changes to the support of the final drive pinion. The "S" Thirty was then retested at Nebraska in 1924 with tractor S2460 and test number 104. Another major change was at S2301 when the transmission was offered as a threespeed.

It was at about serial number S3450 that the last of the true Best Thirty tractors were produced. Many so called "transition" machines were produced that shared similarities between both the Best and Cat Thirty. When production began at Peoria after the 1925 merger, the Thirty tractors produced at that factory were given the serial number PS, starting with PS1. Production of the "S" Thirty ended in 1930 at tractor number S10536 at San Leandro, California.

At serial number PS13286, the split head Thirty was introduced, bearing strong similarity to the Model Thirty-Five. Production of the Model Thirty ended at serial number PS14292 at Peoria in 1931.

The Thirty was available with a number of accessories such as cold weather enclosures, lights, belt pulley and PTO drive units. Later, the rear seat, or tail orchard seat, became an option rather than standard on the Best tractors. The Best color scheme of black with gold trim was kept until the 1925 merger when the gray and red paint color became standard. On December 7 of 1931 the color was changed again to Highway Yellow. It is unclear how many if any of the last Peoria produced Thirty tractors were painted the new color. The Thirty was also available in wide gauge, which is now a highly sought after tractor by collectors. It is also a mystery as to how many wide gauge tractors were produced, but they are quite rare today.

Five years after the S and PS Series Thirty tractors were retired from production, Caterpillar introduced a new tractor that also carried the designation Thirty. This tractor is sometimes labeled the 6G Series Thirty, as "6G" was the prefix to its serial number sequence. In 1937, after Caterpillar had built 875 units, the tractor was redesignated the Caterpillar R4. It remained in production until 1944. We include photographs of the later series Thirty in this book, with the full knowledge that it had little in common with the earlier series.

The 6G Series Thirty was available with either a low-compression, distillate burning or high-compression, gasoline burning 4 cylinder engine, that operated at 1,400 rpm. Its bore and stroke measured 4.25 x 5.50 inches. Its five speed transmission offered operating speeds ranging between 1.7 and 5.4 mph. The low-compression, distillate version developed a maximum 34.13 brake and 26.71 drawbar horsepower; and a maximum 6,120 lbs drawbar pull. The high-compression, gasoline version developed a maximum 37.81 brake and 30.99 drawbar horsepower; and a maximum 7,211 lbs drawbar pull.

The R4 was a re-designation of the 6G Thirty tractor. It was also available in wide or narrow gauge. Many R4 tractors that have been found reflect some type of government affiliation. It seems the R4, like other "R" series tractors, was very popular with the various branches of the government and armed services.

At the end of production, 874 tractors carried the Thirty name and 5,383 carried the R4 name, all with the 6G serial number.

This Book is dedicated to the memory of George Koshak.

Bob LaVoie
October, 1999

Best Thirty equipped with beet shoes pulling Killefer beet plow. Rio Vista, California, September 1928.

Best Thirty pulling Disc plow.

Side view of Best Thirty ripping in a California orchard.

Best Thirty blade grading.

Canopy equipped Best Thirty pulling scrapers.

Best Thirty.

Best Thirty pulling three scrapers. Fordson tractor shown at rear.

Best Thirty.

Pulling a road grader to build a new road.

Best Thirty.

Best Thirty showing the decal style and placement on this near new machine.

Best Thirty works with a road grader as a Best Sixty follows.

Best Thirty numbers 2397-S, 2393-S and 2374-S await shipment from factory.

Best Thirty.

Best Thirty with solid cast radiator sides.

Best Thirty and road grader.

Best Thirty.

Best Thirty with road grader.

Best Thirty with grader. This shows a good view of the early manifold.

Best Thirty with canopy moving a large rock.

Best Thirty with rollover tumble bug-type scraper.

Best Thirty with street pads.

Rare photo of top seat Best Thirty with cab and lights.

Early Caterpillar Thirty with rear mounted seat. At this point, the rear seat was an option rather than standard as on the Best Thirty. Also notice the canopy with the retractable curtains rolled up underneath.

Thirty with top mounted seat introduced in 1926. This machine also has the larger track adjusting springs.

Standard Thirty equipped with optional lighting group.

Caterpillar Thirty equipped for winter work. December 1929.

Thirty equipped with cruiser type canopy top, cold weather enclosures and tow hook. This machine also has front and rear lights and pan style street shoes.

Low level view of the undulating logo showing proper placement on the upper track shields.

Thirty and Ball scraper. This is a later Thirty equipped with late manifold style and larger track adjuster springs.

Rear seat Thirty with Ball scraper at Redwood City, California.

Early Thirty pulling an Adams grader, preparing shoulders for the Caterpillar Trail, a scenic highway along the Illinois River. August 1928. Note the early fuel type "Y" manifold.

Thirty operating three Euclid scrapers on state highway road work. Loudoun County, Virginia. April 1930.

Ethiopian Emperor Heilie Selassie (bearded), accompanied by his sons, views Caterpillar Thirty and grader at work on the Jhimma Road. Wolisso, Ethiopia. November 1932. These machines are equipped with optional engine enclosures and hood.

Two Caterpillar Thirty tractors pulling a Sixty L.W. grader fitted with two 3 foot moldboard extensions. March 1931.

Early Thirty with three Baker Maney self-loading scrapers working on approach to bridge. Notice the early manifold and air cleaner locations.

This machine is a good example of the early manifold and air cleaner style. Notice the pre-heater tube connected to the air cleaner, carburetor and exhaust manifold. This tractor is also equipped with a rear PTO for operating the grader. It also has a Holt style bucket seat.

Caterpillar Sidehill Thirty with Killefer double disc. Vacaville, California. February 1929.

Discing with Caterpillar Thirty. August 1929.

Rear Seat Hill Special Caterpillar Thirty discing in orchard. Watsonville, California. April 1929.

Rear seat Thirty furrowing for irrigation on a date ranch. Indio, California. September 1930.

Eighteen Caterpillar Thirty tractors and combines working in wheat field. The owner of 40,000 acres operated 36 Caterpillar combines. Hays, Kansas. August 1931.

Thirty and Holt combine. Notice spark arrestor on the exhaust pipe.

Rear seat early style Thirty pulling Holt harvester.

Thirty equipped with lighting equipment and Holt combine. Faulkton, South Dakota. October 1928.

Caterpillar Thirty pulling Caterpillar number 3 combine in Lubbock, Texas grain field. July 1931.

Early rear seat Thirty pulling three wagon loads of baled hay. Alviso, California. August 1928.

Rear seat and Holt Western combine.

Early Thirty equipped with street pads and tube type radiator, pulling a Holt harvester.

Caterpillar Thirty pulling 1.5 yard tumble bug scraper. Browns Valley, California. Notice the rear stationary belt pulley drive.

Rear seat Thirty logging in Oregon woods near Klamath Falls.

Caterpillar Thirty with high wheeler, Klamath Indian Reservation. Oregon 1928.

Caterpillar Thirty and Fifteen at the entrance of the Hollow Log Garage. Sequoia National Park. December 1931.

Cold weather equipped Thirty pulling 20 cords of pulp wood. Port Arthur, Ontario. 1929. Notice the skeleton type ice shoes and skid pan.

Thirty equipped with highway trailer reel cart and winch. Ashland, Virginia. June 1930. These machines have also been fitted with street pads and extended track roller frames.

Thirty with Ateco hydraulic scraper working in sand dunes of San Francisco. April 1930.

Thirtys and wagons working under 1.25 yard gasoline shovel. One Thirty equipped with hand operated bulldozer, the other equipped with Willamette single drum winch. Toledo, Ohio. November 1927.

One Thirty, two 5 Tons and a Sixty at work grading Fort Wayne Municipal Airport. November 1928.

Thirty equipped with bulldozer terracing near Washington Monument. January 1932.

Steam shovel loading wagons pulled by a Caterpillar Thirty.

Cold weather equipped Thirty with Killefer loader operating in brickworks. Chicago, Illinois.

Thirty with bulldozer used to smooth off coal after it was unburied. March 1931.

Early Thirty with Euclid wagon at work on the new stadium at the University of Virginia. Charlotteville. October 1930.

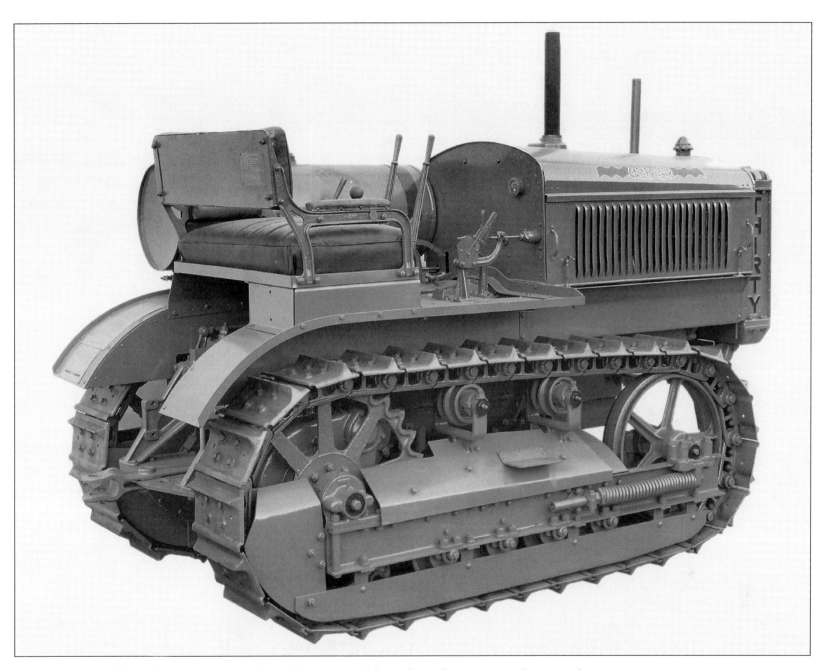
Early Caterpillar Thirty equipped with optional hood and engine side panels.

Thirty with extended track roller frame, 5-roller under carriage and street pads.

This belt drive unit is set up to accommodate a rear seat, which would be attached to the plate on top of the housing.

Notice the serial number tag and patent tag placement.

Late Thirty with rear seat.

Caterpillar Thirty with fully enclosed cab.

Caterpillar Thirty with factory cab. January 1929.

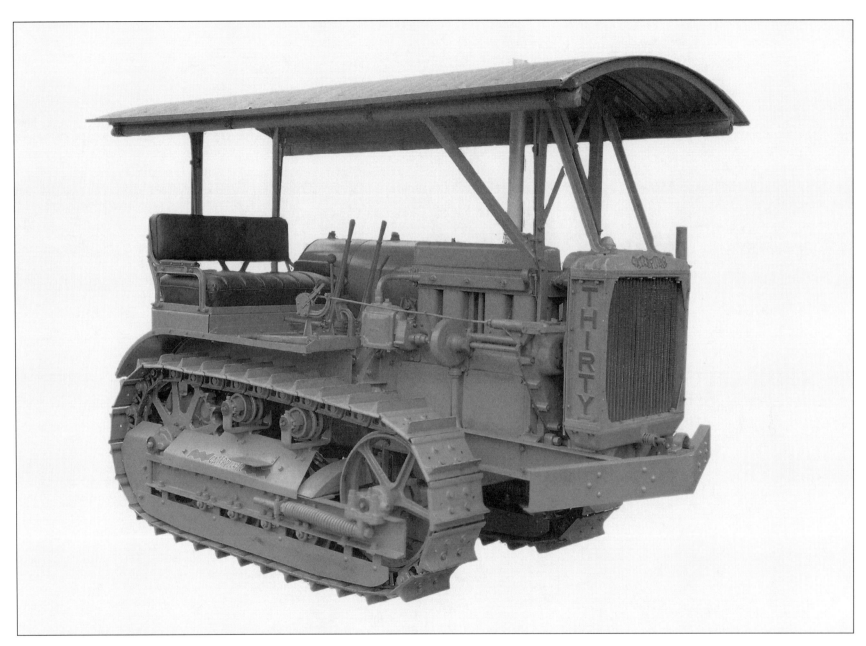

Caterpillar Thirty with optional bumper and canopy.

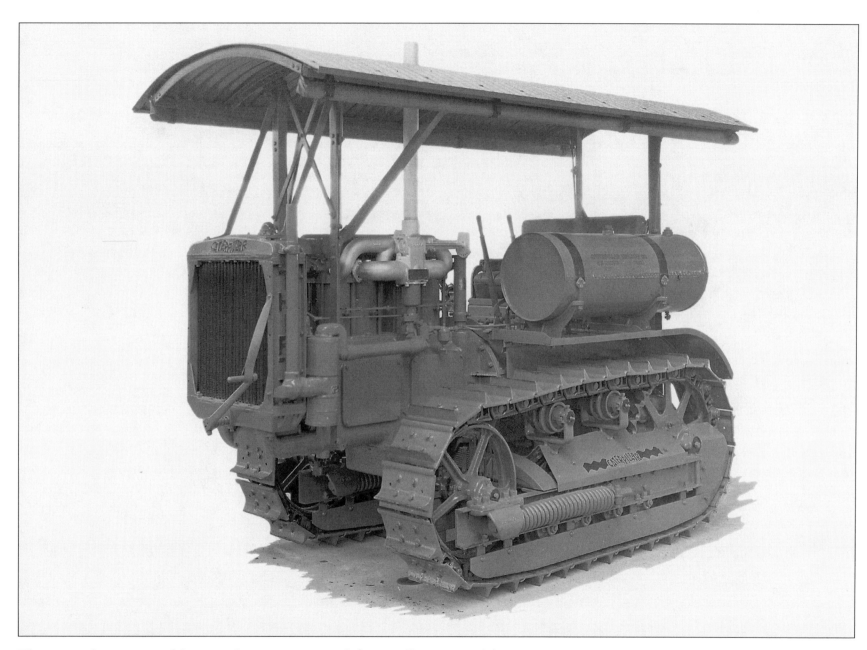

Thirty with improved heavy duty spring and front idler assemblies.

Thirty with Willamette single drum towing winch manufactured in Portland, Oregon.

Cold weather Thirty with snow plow and skeleton ice tracks.

Wide gauge Caterpillar Thirty equipped with Cardwell All-Steel side boom drawing nine strands of one inch copper cable.

Wide gauge model Thirty and rotary scraper working on highway. Bad Lands National Park, South Dakota. August 1931. Notice the lighting equipment and generator.

Wide gauge Thirty and Hi-Way Service scarifier ripping up old road, preparatory to resurfacing. August 1931.

Discing bull thistle with Caterpillar wide gauge Thirty and Dinuba Steel Products disc. July 1928.

Caterpillar wide gauge Thirty with wide angle iron swamp type pads pulling four Athey wagon loads of sugar cane. Lake Okeechobee, Florida. February 1933.

Two wide gauge Caterpillar Thirty tractors pulling 16 foot International Harvester combines in 1400 acre wheat field. Walla Walla, Washington. August 1928.

Wide gauge Thirty and Case combine. August 1922. Also notice lighting equipment.

Wide gauge Thirty pulling two 10 ton wagons through soft fields on a Sherman County, Oregon wheat ranch. August 1930.

Wide gauge Thirty with W-K-M side boom unloading 22 inch, 20 foot pipe. Stinnett, Texas. July 1930.

Two wide gauge Thirty tractors cold bending high carbon 12 inch electric welded oil pipe, both equipped with All-Steel side booms. East Texas. September 1931.

Wide gauge Thirty pulling a 9.75 foot disc. Santa Maria, California. May 1936.

Wide gauge Caterpillar Thirty. 1937.

Thirty Orchard Model with top seat. Notice model designation at radiator side bottom on plate.

Caterpillar Thirty engine as used in 6G series.

Butane Model Thirty.

Butane Thirty (foreground) and Caterpillar RD-6.

Thirty operator controls and instruments.

Thirty with Anthony Model B multiple tool shovel. June 1937.

Wide gauge Thirty with Trackson side boom and winch moving tank sections for a new refinery. Lovell, Wyoming. August 1937.

Trackson high-shovel mounted on Caterpillar Thirty. Milwaukee, Wisconsin. 1937.

Excavating and loading with Thirty and Anthony loader. Peoria, Illinois. July 1937.

Thirty with LeTourneau bulldozer. Oklahoma City, Oklahoma. April 1937.

Wrecking building with Caterpillar Thirty equipped with 4-drum, power controlled unit and LeTourneau tractor crane. Shown lifting out steel girders. Peoria, Illinois. March 1936.

Thirty with Trackson Loader digging a basement. East Peoria, Illinois. November 1937.

Grading shoulders on a California state highway project with Caterpillar Thirty and number 33 grader. March 1937. Notice the factory canopy and louvered side curtains.

Wide gauge Thirty and Killefer rotary scraper tearing down levees for the Leslie Salt Company. Newark, California. January 1938. Notice the bolt on street pads.

Thirty skidding logs near Lisbon, Ohio. May 1936.

Wide gauge 6G series and dozer.

Thirty pulling John Deere 2-bottom, 2-way plow in alfalfa ground and gumbo soil. Fort Morgan, Colorado. April 1936.

Caterpillar Thirty and Caterpillar number 2 terracer at work on the farm of Aetna Life Insurance Company. Lexington, Oklahoma. February 1936.

Thirty pulling 3-bottom plow in old alfalfa field. Oxnard, California. February 1936.

Thirty and hop plow making 8 foot rows between hop vines. The men riding the plows guide outside cutters close to vines. Albany, Oregon. April 1937.

Wide gauge Thirty pulling a 12.5 foot disc and three 6 foot Meeker harrows. Bridgeton, New Jersey. July 1936.

Thirty and number 22 terracer. Hickman County, Kentucky. April 1937.

Thirty pulling 20 foot Rumely combine. Stafford County, Kansas. July 1937.

Caterpillar Thirty pulling Model 36 combine.

Thirty pulling Killefer 8 foot disc in apricot orchard. San Martin, California. April 1936.

Thirty pulling 9.5 foot disc in pear orchard. Agnew, California. April 1936. Notice the front mounted furrow or check breakers.

"Caterpillar" R4 Tractor

Specifications of "Caterpillar" R4 Tractor

CAPACITY: The following are maximum horsepowers at sea level, and are taken from Nebraska Tractor Tests Nos. 271 and 272:	On Gasoline	On "Tractor Fuels"
Drawbar horsepower	35.33	32.39
Belt horsepower	40.83	37.97

The following are the observed drawbar pull, as reported in Nebraska Tractor Tests Nos. 271 and 272:

Drawbar pull:		
First	7,211 lbs.	6,120 lbs.
Second	5,186 lbs.	4,264 lbs.
Third	4,105 lbs.	3,642 lbs.
Fourth	3,147 lbs.	2,536 lbs.
Fifth	2,045 lbs.	1,680 lbs.

The following calculated values for maximum drawbar pull are based on the observed drawbar pull shown above. When slowed down by overload, "Caterpillar" engines develop a considerably greater turning effort at the flywheel (torque), which results in greater drawbar pull at reduced travel speed:

Drawbar pull, maximum:		
First	7,932 lbs.	6,732 lbs.
Second	5,705 lbs.	4,690 lbs.
Third	4,516 lbs.	4,006 lbs.
Fourth	3,462 lbs.	2,790 lbs.
Fifth	2,250 lbs.	1,848 lbs.

Speeds in M.P.H. at full load governed engine R.P.M.:

First (150 ft./min.)	1.7	1.7
Second (211 ft./min.)	2.4	2.4
Third (264 ft./min.)	3.0	3.0
Fourth (325 ft./min.)	3.7	3.7
Fifth (475 ft./min.)	5.4	5.4
Reverse (167 ft./min.)	1.9	1.9

Engine—four-cycle, water-cooled:

	Gasoline	Tractor Fuels
Fuel		
Number of cylinders	4	
Bore and stroke	4¼"x5½"	
Piston displacement	312 cu. in.	
R.P.M.—governed at full load	1,400	
Piston speed	1,283 F.P.M.	
R.P.M. at maximum drawbar pull (point of maximum torque)	900	
N.A.C.C. horsepower rating for tax purposes	28.9	
Lubrication	Force Feed	

Crankshaft:
- Number of main bearings: 5
- Diameter of main bearings: 3"
- Total area main bearing surface: 89.5 sq. in.

Length of tracks on ground (center drive sprocket to center front idler)	5'-1⅛"
Area ground contact (with 13" track shoes)	1,589 sq. in.
Over-all:	
Length	10'-9"
Height (measured from tip of grouser of standard track shoe to highest point exclusive of exhaust pipe and air cleaner inlet screen)	5'-⅝"
Ground clearance (measured from lower face of standard track shoe)	11⅛"
Height drawbar above ground (measured from lower face of standard track shoe)	13⅞"
Lateral movement drawbar (measured at pin)	21"
Track:	
Width of standard track shoe	13"
Height of grouser (measured from upper face of standard track shoe)	2"
Diameter of track shoe bolts	½"
Diameter of track pins	1¾₁₆"
Diameter of track pin bushings	2"
Steering	†
Number friction surfaces in each steering clutch	16
Transmission	‡
Capacities:	
Cooling system, in U.S. Standard gallons	11
Lubricating system:	
Crankcase, in quarts	14
Transmission case, in quarts	20
Final drive case (each), in quarts	7
Fuel tank, in U.S. Standard gallons	32

	60" Gauge	44" Gauge
Over-all width	6'-6"	5'-2"
Weight, shipping (approx.)	9,370 lbs.	9,086 lbs.

†Each track controlled by slow speed, heavy duty dry multiple disc clutch and contracting band brake.

‡Power transmitted through dry type flywheel clutch to selective type change speed gear set.

CATERPILLAR TRACTOR CO., PEORIA, ILLINOIS
MANUFACTURER OF DIESEL ENGINES · TRACK-TYPE TRACTORS · ROAD MACHINERY

TRI-STATE EQUIPMENT CO.

500 E. Overland Avenue

EL PASO, TEXAS

PECOS, TEXAS

FORM 1823-49H   PRINTED IN U.S.A.

R4 with LeTourneau dozer, circa 1950.

More Titles from Iconografix:

AMERICAN CULTURE
AMERICAN SERVICE STATIONS 1935-1943
    ISBN 1-882256-27-1
COCA-COLA: A HISTORY IN PHOTOGRAPHS 1930-1969
    ISBN 1-882256-46-8
COCA-COLA: ITS VEHICLES IN PHOTOGRAPHS 1930-1969
    ISBN 1-882256-47-6
PHILLIPS 66 1945-1954   ISBN 1-882256-42-5

AUTOMOTIVE
CADILLAC 1948-1964   ISBN 1-882256-83-2
CORVETTE PROTOTYPES & SHOW CARS
    ISBN 1-882256-77-8
EARLY FORD V-8S 1932-1942   ISBN 1-882256-97-2
FERRARI PININFARINA 1952-1996 ISBN 1-882256-65-4
IMPERIAL 1955-1963   ISBN 1-882256-22-0
IMPERIAL 1964-1968   ISBN 1-882256-23-9
LINCOLN MOTOR CARS 1920-1942 ISBN 1-882256-57-3
LINCOLN MOTOR CARS 1946-1960 ISBN 1-882256-58-1
PACKARD MOTOR CARS 1935-1942 ISBN 1-882256-44-1
PACKARD MOTOR CARS 1946-1958 ISBN 1-882256-45-X
PLYMOUTH COMMERCIAL VEHICLES
    ISBN 1-58388-004-6
PONTIAC DREAM CARS, SHOW CARS & PROTOTYPES 1928-1998   ISBN 1-882256-93-X
PONTIAC FIREBIRD TRANS-AM 1969-1999
    ISBN 1-882256-95-6
PORSCHE 356 1948-1965   ISBN 1-882256-85-9
STUDEBAKER 1933-1942   ISBN 1-882256-24-7
STUDEBAKER 1946-1958   ISBN 1-882256-25-5

EMERGENCY VEHICLES
AMERICAN LAFRANCE 700 SERIES 1945-1952
    ISBN 1-882256-90-5
AMERICAN LAFRANCE 700&800 SERIES 1953-1958
    ISBN 1-882256-91-3
AMERICAN LAFRANCE 900 SERIES 1958-1964
    ISBN 1-58388-002-X
CLASSIC AMERICAN AMBULANCES 1900-1979
    ISBN 1-882256-94-8
FIRE CHIEF CARS 1900-1997 ISBN 1-882256-87-5
MACK® MODEL B FIRE TRUCKS 1954-1966*
    ISBN 1-882256-62-X
MACK MODEL CF FIRE TRUCKS 1967-1981*
    ISBN 1-882256-63-8
MACK MODEL L FIRE TRUCKS 1940-1954*
    ISBN 1-882256-86-7
SEAGRAVE 70TH ANNIVERSARY SERIES
    ISBN 1-58388-001-1
VOLUNTEER & RURAL FIRE APPARATUS
    ISBN 1-58388-005-4

PUBLIC TRANSIT
THE GENERAL MOTORS NEW LOOK BUS
    ISBN 1-58388-007-0

RACING
GT40   ISBN 1-882256-64-6
JUAN MANUEL FANGIO WORLD CHAMPION DRIVER SERIES
    ISBN 1-58388-008-9
LE MANS 1950: THE BRIGGS CUNNINGHAM CAMPAIGN   ISBN 1-882256-21-2
LOLA RACE CARS 1962-1990 ISBN 1-882256-73-5
LOTUS RACE CARS 1961-1994 ISBN 1-882256-84-0
MARIO ANDRETTI WORLD CHAMPION DRIVER SERIES
    ISBN 1-58388-009-7
MCLAREN RACE CARS 1965-1996 ISBN 1-882256-74-3
SEBRING 12-HOUR RACE 1970 ISBN 1-882256-20-4
VANDERBILT CUP RACE 1936 & 1937
    ISBN 1-882256-66-2
WILLIAMS 1969-1999 30 YEARS OF GRAND PRIX RACING
    ISBN 1-58388-000-3

RAILWAYS
CHICAGO, ST. PAUL, MINNEAPOLIS & OMAHA RAILWAY 1880-1940   ISBN 1-882256-67-0
CHICAGO&NORTH WESTERN RAILWAY 1975-1995
    ISBN 1-882256-76-X
GREAT NORTHERN RAILWAY 1945-1970
    ISBN 1-882256-56-5
GREAT NORTHERN RAILWAY 1945-1970 VOLUME 2
    ISBN 1-882256-79-4
MILWAUKEE ROAD 1850-1960 ISBN 1-882256-61-1
SOO LINE 1975-1992   ISBN 1-882256-68-9
TRAINS OF THE TWIN PORTS, DULUTH-SUPERIOR IN THE 1950s   ISBN 1-58388-003-8
WISCONSIN CENTRAL LIMITED 1987-1996
    ISBN 1-882256-75-1
WISCONSIN CENTRAL RAILWAY 1871-1909
    ISBN 1-882256-78-6

TRUCKS
BEVERAGE TRUCKS 1910-1975 ISBN 1-882256-60-3
BROCKWAY TRUCKS 1948-1961* ISBN 1-882256-55-7
DODGE PICKUPS 1939-1978 ISBN 1-882256-82-4
DODGE POWER WAGONS 1940-1980 ISBN 1-882256-89-1
DODGE TRUCKS 1929-1947 ISBN 1-882256-36-0
DODGE TRUCKS 1948-1960 ISBN 1-882256-37-9
LOGGING TRUCKS 1915-1970 ISBN 1-882256-59-X
MACK MODEL AB*   ISBN 1-882256-18-2
MACK AP SUPER-DUTY TRUCKS 1926-1938*
    ISBN 1-882256-54-9
MACK MODEL B 1953-1966 VOL 1* ISBN 1-882256-19-0
MACK MODEL B 1953-1966 VOL 2* ISBN 1-882256-34-4
MACK EB-EC-ED-EE-EF-EG-DE 1936-1951*
    ISBN 1-882256-29-8
MACK EH-EJ-EM-EQ-ER-ES 1936-1950*
    ISBN 1-882256-39-5
MACK FC-FCSW-NW 1936-1947* ISBN 1-882256-28-X
MACK FG-FH-FJ-FK-FN-FP-FT-FW 1937-1950*
    ISBN 1-882256-35-2
MACK LF-LH-LJ-LM-LT 1940-1956* ISBN 1-882256-38-7
MACK TRUCKS PHOTO GALLERY* ISBN 1-882256-88-3
NEW CAR CARRIERS 1910-1998 ISBN 1-882256-98-0
STUDEBAKER TRUCKS 1927-1940 ISBN 1-882256-40-9
STUDEBAKER TRUCKS 1941-1964 ISBN 1-882256-41-7
WHITE TRUCKS 1900-1937 ISBN 1-882256-80-8

TRACTORS & CONSTRUCTION EQUIPMENT
CASE TRACTORS 1912-1959 ISBN 1-882256-32-8
CATERPILLAR THIRTY 2ND EDITION INCLUDING BEST THIRTY 6G THIRTY & R-4 ISBN 1-58388-006-2
CATERPILLAR D-2 & R-2 ISBN 1-882256-99-9
CATERPILLAR D-8 1933-1974 INCLUDING DIESEL 75
    ISBN 1-882256-96-4
CATERPILLAR MILITARY TRACTORS VOLUME 1
    ISBN 1-882256-16-6
CATERPILLAR MILITARY TRACTORS VOLUME 2
    ISBN 1-882256-17-4
CATERPILLAR SIXTY ISBN 1-882256-05-0
CATERPILLAR PHOTO GALLERY ISBN 1-882256-70-0
CLETRAC AND OLIVER CRAWLERS ISBN 1-882256-43-3
ERIE SHOVEL ISBN 1-882256-69-7
FARMALL CUB ISBN 1-882256-71-9
FARMALL F– SERIES ISBN 1-882256-02-6
FARMALL MODEL H ISBN 1-882256-03-4
FARMALL MODEL M ISBN 1-882256-15-8
FARMALL REGULAR ISBN 1-882256-14-X
FARMALL SUPER SERIES ISBN 1-882256-49-2
FORDSON 1917-1928 ISBN 1-882256-33-6
HART-PARR ISBN 1-882256-08-5
HOLT TRACTORS ISBN 1-882256-10-7
INTERNATIONAL TRACTRACTOR ISBN 1-882256-48-4
INTERNATIONAL TD CRAWLERS 1933-1962
    ISBN 1-882256-72-7
JOHN DEERE MODEL A ISBN 1-882256-12-3
JOHN DEERE MODEL B ISBN 1-882256-01-8
JOHN DEERE MODEL D ISBN 1-882256-00-X
JOHN DEERE 30 SERIES ISBN 1-882256-13-1
MINNEAPOLIS-MOLINE U-SERIES ISBN 1-882256-07-7
OLIVER TRACTORS ISBN 1-882256-09-3
RUSSELL GRADERS ISBN 1-882256-11-5
TWIN CITY TRACTOR ISBN 1-882256-06-9

*This product is sold under license from Mack Trucks, Inc. Mack is a registered Trademark of Mack Trucks, Inc. All rights reserved.

All Iconografix books are available from direct mail specialty book dealers and bookstores worldwide, or can be ordered from the publisher. For book trade and distribution information or to add your name to our mailing list contact

Iconografix
PO Box 446
Hudson, Wisconsin, 54016

Telephone: (715) 381-9755
(800) 289-3504 (USA)
Fax: (715) 381-9756

MORE GREAT BOOKS FROM ICONOGRAFIX

CATERPILLAR SIXTY Photo Archive
ISBN 1-882256-05-0

CATERPILLAR MILITARY TRACTORS VOL 1 Photo Archive
ISBN 1-882256-16-6

CATERPILLAR D-8 1933-1974 Photo Archive
ISBN 1-882256-96-4

CATERPILLAR MILITARY TRACTORS VOL 2 Photo Archive
ISBN 1-882256-17-4

CATERPILLAR D-2 & R-2 Photo Arcive
ISBN 1-882256-99-9

CATERPILLAR Photo Gallery
ISBN 1-882256-70-0

HOLT TRACTORS Photo Archive
ISBN 1-882256-10-7

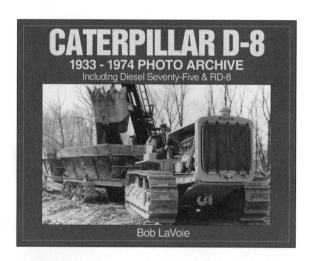

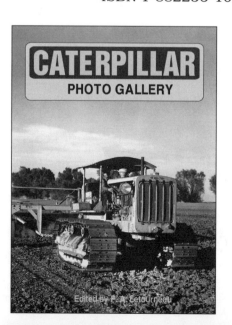
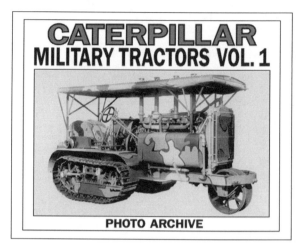
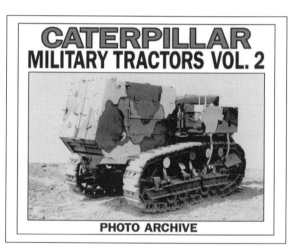
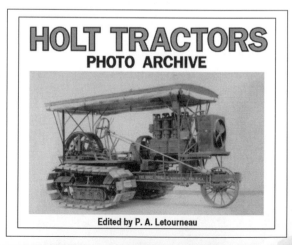